25 YEARS OF THE CLASS 66

John Jackson

AMBERLEY

First published 2024

Amberley Publishing
The Hill, Stroud
Gloucestershire, GL5 4EP

www.amberley-books.com

ISBN 978 1 3981 0898 1 (print)
ISBN 978 1 3981 0899 8 (ebook)

British Library Cataloguing in Publication Data.
A catalogue record for this book is available from the British Library.

Origination by Amberley Publishing.
Printed in the UK.

Contents

Introduction

For many railway enthusiasts, including myself, the first glimpse of the newly delivered initial batch of Class 66 locomotives was in August 1998. The end of that month saw an open day at Toton depot, hosted by English, Welsh & Scottish Railways. On display were the first four deliveries of the class, numbers 66001, 66003, 66004 and 66005. Sister loco No. 66002 had been undergoing further testing and consequently did not arrive in the UK until the following year.

The delivery of these four locos was the culmination of a planning process that had started two years earlier. British Rail's freight operations had been privatised in the mid-1990s, with Wisconsin Central Transportation Services successfully buying a number of operating companies which were to be rebranded as English, Welsh & Scottish Railways (EWS). Their controller, Ed Burkhardt, had appraised the existing motive power now under his company's ownership, consisting of around 1,600 locos, with an average age of thirty years. In addition, a number of these inherited locos had already been stripped for spares. He concluded that there was a need to both reduce overall running costs and improve reliability. A review of the Class 59 locomotives already in use in the UK suggested that both of these objectives had been achieved in the operation of this fleet of fifteen locos.

The concept of the UK Class 66 was born, and his company committed to an order for a total of 250 locomotives, modelled, in part, on the Class 59. These locos were to be built by General Motors at the plant of their Electro-Motive division in London, Ontario, Canada. By way of comparison, the fuel tank capacity of this new loco class was approximately double that of the Class 47 locos, for example, which constituted one of the biggest classes of diesel locomotives inherited from the pre-privatisation era.

These new locomotives were to be numbered from 66001 to 66250 inclusive. The first delivery was loco No. 66001, which arrived via Immingham Docks on South Humberside in June 1998. Regular deliveries through Newport Docks, in South Wales, then followed, with the entire order being completed by the end of 2001. Equally impressive was the instant availability of these locos on arrival such that they could be put into service almost immediately.

It did not take the enthusiast fraternity long to give these EWS Class 66s the nickname 'sheds', with their upturned roof resembling that of a shed!

In 2007, EWS was sold to German-based Deutsche Bahn, and rebranded in the UK as DB Schenker. Today, it forms a part of the Europe-wide transport operations of Deutsche Bahn and has been restyled as DB Cargo.

Back in the 1990s, the arrival of EWS's Class 66s did not go unnoticed amongst other railfreight operators and, in particular, Freightliner. They were to place an initial order for five locomotives in 1998 numbered from 66501 upwards. This was to be followed in 2000 by Freightliner's creation of a sub class of 66/6 locomotives, with a lower gear ratio. This enabled heavier trains to be hauled, albeit at a reduced speed.

Since that Freightliner commitment to the General Motors' loco, GB Railfreight, Direct Rail Services and Colas Rail have all placed orders for these, ensuring the class is found right across the UK rail network. Today, the class can be found on duties ranging from the china clay traffic in Cornwall to both passenger sleeper and freight container services in the Scottish Highlands.

Their success has ensured that a total in excess of 500 locos have seen service in the United Kingdom. The majority of these have been delivered as newly built locos, in both Canada and the United States.

An industrial dispute at the plant in London, Ontario, Canada, led to the workers there being 'locked out' and the Canadian operation was closed down by its owners who threatened to move any future production to the USA.

This closure coincided with a general downturn in freight traffic, notably in Europe. This downturn resulted in GB Railfreight (GBRf) seeking additional locomotives through second-hand acquisition from a number of operators, rather than through new-build locos in North America.

By 2013, GBRf's continued growth, however, led to a further order for new Class 66 locos to be built by Electro-Motive Diesel's manufacturing base at Muncie, Indiana, USA. This move was formally recognised by the naming of their loco No. 66752, *The Hoosier State*, the US nickname for the state of Indiana. Increasingly stringent European emissions regulations meant that their loco, numbered 66779, was to be the last to be newly built. It led to the loco being appropriately named *Evening Star*. It arrived in the UK in February 2016, with the loco entirely 'wrapped' ensuring no premature viewing of its livery and nameplates.

By contrast to GBRf's continued acquisition of locos bringing its UK Class 66 total to the 100 mark, DB Cargo had by now relocated a number of its original build of 250 locos for continental work. A fleet of these locos continues to operate in both France and Poland, leaving the remaining 160 or so available in the UK. A few of these machines have languished at the company's Toton depot having been out of revenue-earning action for months, and in a few cases years. A few French examples have also been returned to Toton in the last few months, potentially for a return to UK duties.

Likewise, Freightliner has also despatched a smaller number operating as a fleet in Poland. Additionally, some locos have seen transfer between the different freight operators in the UK. Space precludes the covering of all such moves in any detail here, but suffice to say that a total of over 400 Class 66 locos are operated by the UK's five main freight operators at the time of compiling these notes.

Despite the large number of locos in the class, serious accidents and damage have, thankfully, been few and far between. DB Cargo's No. 66048 and GBRf's No. 66734 both suffered accident damage in Scotland and have subsequently been withdrawn. Likewise, Freightliner's No. 66521 was involved in a fatal accident at Great Heck, near Selby, North Yorkshire, in 2001 and subsequently written off. Their sister loco, No. 66526, carries the name of the driver, Steve (George) Dunn, who, sadly, lost his life in the accident. GBRf's original No. 66734 has now been reallocated to a more recent addition to its fleet.

In the pages that follow, we take a detailed look at the class pioneers for the main operators, followed by a cross section of the operations these locos have found themselves employed on for a quarter of a century. These are arranged broadly in numeric order, company by company, resulting in almost 200, or nearly half the combined fleet, making an appearance in this publication.

Finally, I hope you enjoy browsing through these pages as much as I have enjoyed compiling them. It has reminded me of those days, full of promise, back in 1998 when EWS unveiled the locos it promised to be railfreight's 'new dawn'.

John Jackson

The Class Pioneers

The UK delivery of the Class 66 locomotives commenced in 1998. For many enthusiasts, myself included, the first glimpse of these new locos was at the open day held at Toton depot in August that year, hosted by English, Welsh & Scottish Railways (EWS). Their pioneer, No. 66001, was amongst the locos on display.

Second-numbered No. 66002 was still away on test at the time, leaving three other pioneers to be on display at Toton that weekend, namely Nos 66003, 66004 and 66005. In this photo, the gathered enthusiasts get the chance to get close to No. 66003.

Freightliner took delivery of its first Class 66 locos a year later. One of their pioneers, No. 66502, was amongst the exhibits at the open day at Crewe Works in May 2000. It is seen here on the works' traverser.

The third railfreight operator to enter the Class 66 market was GB Railfreight (GBRf). In 2000, they were awarded a contract to operate trains on behalf of Railtrack, the then infrastructure owner. An order for seven Class 66s was placed the same year, to be numbered 66701 to 66707 inclusive. On 22 June 2001, No. 66701, then named *Railtrack National Logistics*, heads north through Peterborough on one of the Railtrack workings.

EWS/DB Cargo

EWS were the first freight operator to put its Class 66 locomotives into service in the UK. On the following pages we take a look at a cross section of these workings under both EWS and DB Cargo ownership. On 3 July 2018, No. 66003 passes Oakley, north of Bedford, on a loaded aggregates working from Mountsorrel Quarry, in Leicestershire, to Elstow, south of Bedford.

DB Cargo chose No. 66004 to carry a livery promoting the company's green credentials. With 'I am a climate hero' vinyls, the loco passes Bedford on 22 December 2021 on a rake of empty aggregate hoppers returning from Radlett to Mountsorrel.

Another livery variation has been applied to seven DB Cargo Class 66 locos. In 2019, DB Cargo and Maritime Transport announced their joint launch of Maritime Intermodal, incorporating DB Cargo running Maritime Transport's container services out of both Felixstowe and Southampton. On 28 April 2021, No. 66005, named *Maritime Intermodal One*, passes Manea, Cambridgeshire, on a working from Wakefield to Felixstowe.

The company has a long association with steel traffic, particularly to and from South Wales. On 13 November 2017, No. 66007 passes through Severn Tunnel Junction with a return working from Cocklebury Yard, Swindon, to the yard at Margam, in South Wales.

Another South Wales working on 15 July 2015 sees No. 66011 passing through Cardiff's Central station on a rake of vans. The 'shed' is on a local working from Bridgend to the yard at Ebbw Junction, Newport.

DB Cargo has seen the regular flow of coal traffic replaced, in part, by the movement of biomass in recent years. On 6 July 2018, No. 66012 heads through New Barnetby on a working from the docks at Immingham, South Humberside, to the power station at Drax, North Yorkshire.

In 2009, EWS was rebranded as DB Schenker and started repainting its loco fleet in its new red livery. One such example, No. 66017, is seen taking the curve as it heads north through Leamington Spa. It is hauling one of the company's regular container workings from Southampton Docks. This particular service is destined for the inland terminal at Birch Coppice, near Tamworth.

An early duty for EWS Class 66s was the regular delivery of concrete blocks on behalf of Plasmor Ltd. The company uses rail for delivery from its base at Great Heck, Yorkshire, to both Biggleswade, Bedfordshire, and Bow, East London. On 22 June 2001 No. 66019 waits at Peterborough on its northbound journey returning the empty wagons to Yorkshire.

On 25 June 2020, No. 66023 is seen on a longer-distance container service from Southampton passing Water Orton, in the Birmingham suburbs. This particular working is destined for Masborough, near Rotherham.

One of the few remaining freight services in the West Country is the flow of locally mined china clay to the docks at Fowey, in Cornwall, for export. On 2 June 2016, No. 66027 approaches St Blazey on a loaded rake from Goonbarrow to Fowey.

DB Cargo's Class 66 loco fleet has been regularly used on weekend engineering work on behalf of Network Rail, often involving light engine convoy movements in order to undertake this work. On Saturday 29 April 2017, for example, No. 66034 leads Nos 66002, 66004 and 66093 on one such move. The quartet are about to enter Eastleigh's East Yard having arrived from Margam, South Wales.

On 19 July 2016, the regular steel flow between Scunthorpe and Eastleigh was in the hands of No. 66037. Usually containing long welded rail for infrastructure use, at present this working is operated by GBRf. The train is seen heading south through Chesterfield.

The approach to Derby station is the location for this view of another long-standing steel flow. On 3 February 2017, No. 66040 is seen with the empty wagons returning from Wolverhampton Steel Terminal to Immingham, South Humberside.

A very different duty fell to No. 66047 on 17 June 2017. It was called upon to assist steam locomotive No. 60009 from the National Railway Museum, York, to Stewarts Lane in Battersea, South London. The pair are passing Tempsford crossing, to the north of Sandy, Bedfordshire.

This view of Saltley depot, in Birmingham, on 18 May 1999 reflects the EWS transition from old to new. Class 66 locomotive No. 66053 is stabled alongside Class 58 loco No. 58030. The Class 66 was newly delivered that year while the Class 58 was built in 1985 and ended its days working in Spain.

On 27 June the same year, two class members are seen in the northern outpost of Inverness's Millburn Yard. At the time, EWS were handling rail deliveries for supermarket chain Safeway, for onward movement of products for the outlets in the north of Scotland. On that day, No. 66060 stands alongside sister loco No. 66119 in the yard.

On 16 June 2021, No. 66061 heads west through Reading on the working of the 'binliners' between Northolt and Severnside. The train is delivering to the Energy from Waste (EfW) power station at Severnside, on the Severn Beach branch, near Bristol.

All operators' Class 66 locomotives have handled infrastructure work on behalf of Railtrack and Network Rail. On 14 June 2003, EWS's No. 66073 stands at the head of a rake of engineering wagons at Rugby.

The Class 66 locos were delivered just in time to be paired with rakes of the outgoing 'merry go round' wagons that had been the mainstay on coal workings since the mid-1960s. On 29 April 1999, No. 66073 heads through Newport, South Wales, on a westbound coal working.

Kellingley Colliery, North Yorkshire, closed in 2015, marking the end of deep-pit coal mining in Britain. DB Cargo Class 66s were subsequently employed in removing spoil from the site. On 19 April 2018, No. 66079 *James Nightall GC* waits to leave the site with a working to Killingholme, South Humberside.

Class 66 No. 66080 heads across the Cambridgeshire Fens on a service destined for the major container port of Felixstowe in Suffolk. The working has left the inland terminal at East Midlands Gateway and is seen at North Fen, near Little Downham.

A significant and welcome additional freight flow commenced in early 2022. Substantial investment by DB Cargo at its Toton base, on the Derbyshire and Nottinghamshire county boundary, has facilitated loading of Toyota cars for export. Their Corolla model, built nearby at Derby, is being exported to France and the Czech Republic, via Dollands Moor and the Channel Tunnel. On 16 March 2022, No. 66082 approaches Kettering on this working.

Another Maritime-liveried Class 66, No. 66090 *Maritime Intermodal Six*, is seen on Friday 8 October 2021, in Tyne Yard. In company with other classmates, it is shunting rakes of loaded wagons in readiness for weekend engineering work.

DB Cargo chose No. 66099 to be branded with Ukraine-related vinyls. It is seen, along with No. 66019, waiting to head north from Doncaster on 9 June 2022. The loco's side is adorned with the blue and yellow Ukraine flag and the words 'We Stand With Ukraine'.

On 24 September 2020, No. 66100 *Armistice 100 1918–2018* is in charge of a rake of vans as it passes Old Linslade, north of Leighton Buzzard, on the West Coast Main Line. The train, worked by the Class 66 from Dollands Moor to Daventry, is carrying imported bottled water.

Curtain-sided wagons form the majority of the consist of this steel working on Saturday 1 July 2017. With No. 66103 in charge, this working from Margam, South Wales, to Dee Marsh, near Shotton in North Wales, was diverted due to weekend engineering work on the Marches Line. It is seen whilst off route as it passes Codsall station.

Newly liveried No. 66105 and its wagonload complement each other on an all-red DB Cargo aggregate working from Dowlow, near Buxton, to Theale, Berkshire. The train is seen passing through Nuneaton station on 30 September 2019.

Uninviting January weather greets No. 66107 as it works south through Tamworth's low-level platforms on 19 January 2017. It is hauling a rake of loaded car containers on a service from Halewood, Merseyside, to Southampton's Docks.

A unique livery has been applied to No. 66109, together with the name Teesport Express. The dark blue colour reflects the working relationship with PD Ports. It coincided with the launch of a second service operated by DB Cargo, this time linking Teesport with Grangemouth, in Central Scotland. The loco is to be found on a different working on 24 May 2021. It passes Colton Junction, near York, on a binliner service from Tees Yard to Knowsley, Merseyside.

DB Cargo's No. 66111 is on the rear of a three-loco convoy, along with No. 66055 and GB Railfreight's No. 66722, on a move between the yards at Bescot Yard, West Midlands, and Toton. The trio are passing Burton on Trent on 23 April 2019.

Several Class 66 locos were regularly employed on Scottish coal workings, particularly on the flow across Central Scotland between Hunterston and Longannet. The imported coal was loaded at the Hunterston terminal at Fairlie, on the Firth of Clyde, and moved to the now closed power station at Longannet, near Kincardine, on the Firth of Forth. On 31 March 2015, No. 66114 takes a loaded train eastwards across Central Scotland and is seen passing through Greenfauld station.

On 20 October 2020, No. 66115 is one of that day's locos employed on aggregate traffic from the quarries around Buxton, in Derbyshire. Having left Briggs Sidings, at Dowlow, the train has reached Great Rocks Junction signal box and the driver is about to hand over the single line token to the signalman. The train will then work to Ashburys, in Manchester.

On 20 June 2013, No. 66116 heads through Walsall station in the West Midlands. It is returning a rake of empty wagons from Wolverhampton Steel Works to Immingham, South Humberside.

The fleet of DB Cargo Class 66s have been regular performers on the steel traffic between Scunthorpe and Ebange, France, to and from the Channel Tunnel freight yard at Dollands Moor in Kent. On 26 April 2014, No. 66117 passes Marholm, north of Peterborough, on the northbound working returning the empty wagons to Scunthorpe.

In 2015, the substantial volume of aggregates delivered by rail from the Mendip quarries at Merehead and Whatley was in the hands of DB Cargo. One such working, hauled by No. 66118, is passing through Southall station on 10 September that year, heading for the yard at Acton, West London.

A wintry scene provides the backdrop as No. 66121 passes through Barnetby, on South Humberside, on 15 January 2013. The train, loaded with imported coal, is making the short journey from Immingham Docks to Scunthorpe.

On 2 October 2019, No. 66128 passes through Kensal Rise, in North London. It is working a rake of aggregate hoppers from Marks Tey, in Suffolk, to Crawley New Yard, in West Sussex.

DB Cargo's No. 66136 was chosen to be adorned with the Yiwu–London Train branding. This was to mark the inaugural container run from Yiwu, Eastern China, to London, which arrived via the Channel Tunnel in January 2017. The so-called Silk Road train had covered 7,500 miles in fourteen days, with the containers transferred twice between wagons during their journey. The loco is seen on 7 February that year passing Leicester on an intermodal to Felixstowe.

Ealing Broadway station in West London is the setting in this photo taken on 24 September 2015. No. 66142 heads east with a rake of hoppers from the nearby Tarmac terminal at Hayes to the Leicestershire quarry at Cliffe Hill.

With the substantial site of Lackenby on the right of this view, No. 66144 approaches Grangetown, on Teesside, on 18 April 2018. The short train is a local working from Skinningrove to Tees Yard, Thornaby.

Back in 2017, the rail-served cement works at Ketton, in Rutland, was a long-standing customer of EWS/DB Cargo. On 2 November 2017, No. 66151 is seen returning a rake of empty wagons from Churchyard Sidings, close to London's St Pancras station, back to Ketton as it passes Harrowden Junction, north of Wellingborough. This working is now in the hands of GB Railfreight.

Another long-standing DB Cargo working involves the movement of steel coils from Margam, in South Wales, to the steelworks at Corby, in Northamptonshire. On 10 October 2014, No. 66152 *Derek Holmes – Railway Operator* heads north through Loughborough station on a loaded working destined for Corby.

Cumbernauld station, to the east of Glasgow, is the location of this coal working across Central Scotland. On 12 June 2013, No. 66158 is seen heading a loaded rake from Hunterston to Longannet Power Station. Longannet was to close in 2016, with these coal workings consigned to history.

Barnetby station sees regular flows of loaded oil trains from the terminals at Humber and Lindsey, routinely hauled by DB Cargo Class 60 locos. Their Class 66s do make appearances, however, as evidenced here on 12 July 2017. A working from Humber to Kingsbury, near Tamworth, passes Barnetby East signal box as No. 66161 takes the curve into Barnetby station.

On 6 August 2020, another of DB Cargo's Maritime-liveried examples is seen passing Whittlesea Signal Box. No. 66162 *Maritime Intermodal Five* heads west on a working from Felixstowe to the terminal at East Midlands Gateway.

On 1 September 2016, No. 66164 heads east through Hayes & Harlington station on a stone working from the quarry at Whatley to St Pancras Churchyard Sidings.

The Electro-Motive depot at Longport, near Stoke-on-Trent, carries out Class 66 locomotive overhauls. On 2 September 2015, No. 66167 stands in the small yard outside the maintenance facility, viewed from passing on the adjacent Crewe to Derby line.

The former EWS coal hoppers are now seeing alternative use on stone workings for DB Cargo. On 6 April 2018, No. 66171 passes Three Horseshoes crossing, near the Cambridgeshire village of Turves. The loco is nearing journey's end on the working from Peak Forest to Ely.

On 18 March 2016, No. 66172 *Paul Melleney* approaches the end of the platform at Tame Bridge Parkway, West Midlands. It is working a short rake of empty steel wagons from the Wolverhampton Steel Terminal back to Masborough, Rotherham.

On 30 May 2017, No. 66175 approaches Par station on a loaded china clay working from Parkandillack, near St Austell. The train will reverse at Lostwithiel before heading down the single-track branch line to Carne Point, Fowey.

On 12 November 2016, a Saturday, No. 66182 passes Stewartby, Bedfordshire, on the Marston Vale branch, with the empty wagons from an earlier loaded working from Peak Forest to Bletchley. The rake will stable for the weekend at Doncaster Belmont Yard, due to siding capacity constraints in the Peak Forest quarry area.

The impressive roof of Carlisle's Citadel station is the backdrop as No. 66192 takes the centre road on 9 April 2015. The infrastructure train was working from Carlisle to Tyne Yard, south of Newcastle.

Back in 2013, this was a familiar sight at one of my favourite locations. On 8 June that year, EWS's No. 66194 passes Hillam Gates, near the village of Monk Fryston, North Yorkshire. It is on a positioning move taking empty coal hoppers from Toton to the nearby yard at Milford Junction.

With the demise of coal traffic, the importance of aggregates traffic to the railfreight sector cannot be overstressed. On 4 August 2018, No. 66221 approaches Bedford on a loaded working from Mountsorrel to Elstow. The photo was taken before this view was obscured as a result of Midland Main Line electrification work.

EWS Class 66 No. 66048 was written off following its derailment at Carrbridge, on the Highland Main Line, in January 2010. This loco aside, the DB Cargo Class 66 UK fleet remains intact, although several examples have not seen revenue service for some time. No. 66230, for example, was fire damaged following a collision with a petrol-powered buggy as it entered the Channel Tunnel freight terminal at Dollands Moor in September 2018. The loco has not returned to traffic and is currently stored at Toton depot. In December 2017, the loco is seen stabled at Hoo Junction yard, in Kent.

Much has changed on our rail network since this photo was taken on 22 May 2001, as No. 66245 is signal checked alongside Warrington Bank Quay station. The 'merry go round' coal hoppers have long been consigned to history, whilst the loco is now operating for DB Cargo in France.

DB Cargo France

In 2005, EWS formed Euro Cargo Rail (ECR) to enable freight operations in France. It was subsequently included in the sale to Deutsche Bahn two years later and renamed DB Cargo France in 2021. A total of sixty-five Class 66 locos were transferred to France from the UK, including No. 66028. It is seen here, on 21 December 2012, passing Chesterfield whilst being dragged by No. 66250 within the consist of a Doncaster to Toton working.

In addition to the internal modifications required to work abroad, cosmetic changes to the locos were necessary. These changes are evident as No. 66218 approaches Peterborough working the northbound empty Plasmor block wagons on 10 September 2012.

These DB Cargo France locos have occasionally returned to the UK depot at Toton for attention. In 2021, however, a total of nine locos were returned to the UK for modification and return to service here. On 18 June 2021, No. 66010 is on the rear of a convoy, also conveying Nos 66190 and 66028, being hauled by No. 66090 in the path of a wagonless Dollands Moor to Scunthorpe. The trio were ultimately destined for Toton depot.

At Toton on 2 September 2021, another French returnee, No. 66032, is seen being shunted around the depot complex. Fifteen former UK-based locomotives are now in operation with DB Cargo Polski. None of these has returned since exportation.

Direct Rail Services

DRS currently has five Class 66 locos on long-term hire from DB Cargo, each of which has been branded in DRS livery. These include No. 66091, seen ahead of No. 66430, as the pair arrive at Nuneaton on 31 August 2021. They are working a rake of infrastructure wagons from Crewe to Cliffe Hill (Stud Farm) for loading at the Leicestershire quarry.

In 2011, DRS acquired five Class 66s, numbered 66301 to 66305, from Fastline, which had gone into administration the previous year. On 12 October 2021, No. 66301 *Kingmoor TMD* crosses Tomatin viaduct, on the Highland Main Line, on the daily Inverness to Mossend working.

Covid-related lockdowns apart, DRS offers the public the chance to view its two main bases, at Crewe and Carlisle, with annual open days, alternating between the two sites. In 2019 it was the turn of Carlisle Kingmoor to host the event. On 20 July, No. 66302 awaits its naming as *Endeavour*.

The company handles infrastructure workings across Soctland on behalf of Network Rail. On 7 October 2017, a Saturday, No. 66303, along with No. 66302 on the rear, works a loaded ballast train from Millerhill Yard, near Edinburgh, through Stirling station. The next day, the train will drop its load on Scotland's Far North Line, close to the remote outpost of Forsinard.

Following the transfer of twenty of its Class 66 locos to Freightliner and GB Railfreight, DRS retained a further fourteen examples, Nos 66421 to 66434. One of these, No. 66422, is seen heading north through Doncaster on 20 June 2018 on an engineers' working from Doncaster to Millerhill.

The Inverness to Mossend container working is seen again, this time on 9 October 2017. Not long after leaving the Highland capital, No. 66423 passes through Carrbridge, having waited for a track machine to clear the single-track section to the north of here.

DRS operates a weekday service between its Daventry hub and the South Wales terminal at Wentloog, Cardiff. On 20 April 2021, No. 66424 passes Water Orton, in the Birmingham suburbs, on the Wales-bound working.

The 2019 DRS open day gave the Carlisle Kingmoor visiting public a three-dimensional view of No. 66426. It is seen here on aerial display inside the depot.

The long-distance Anglo-Scottish container services are the money-spinning aspirations of the major freight operators, with DRS included in the competitive mix. On 11 March 2014, No. 66429 passes through the low-level platforms at Tamworth on a Coatbridge to Daventry working.

In recent years, GB Railfreight (GBRF) has hired in Class 66 locos from DRS to assist with its shunting needs at Dove Holes Quarry, Peak Forest. On 19 October 2020, No. 66433 is seen on these manoeuvres, with GBRf's own No. 66737 in the background.

Freightliner

A batch of ten locomotives moved from DRS to Freightliner, retaining their original numbers 66411 to 66420. Three of these, Nos 66411, 66412 and 66417, were subsequently exported for use by Freightliner in Poland. On 29 September 2016, one of the remaining seven, No. 66414, eases through Ipswich station hauling a Bristol to Felixstowe freightliner.

The Freightliner business was established as a management buyout around the time of rail privatisation in the mid-1990s. The company was eventually acquired by US company Genesee & Wyoming in 2015. The parent company's livery soon started appearing on its fleet of Class 66s, including No. 66415 *You Are Never Alone*, seen here on 15 September 2020. It is passing Fairwood Junction, on the outskirts of Westbury, with a return working from Westbury to Fairwater Yard, near Taunton.

Another Freightliner container train is seen approaching Nuneaton on 9 March 2017. Their Class 66, No. 66418, hauls a freightliner from Lawley Street FLT in Birmingham to the Suffolk Port of Felixstowe.

Sister loco No. 66420 *Patriot in Memory of Fallen Railway Employees* was called upon to rescue Class 70 No. 70008 on 20 April 2021. The cement train, from Earles Sidings, in Derbyshire's Hope Valley, to Walsall, West Midlands, is seen approaching Water Orton. The Class 66 had been added at Toton.

Freightliner's first numbered 66501, now named *Japan 2001*, was delivered back in 1999. Twenty years later, on 29 August 2019, the loco draws into Nuneaton for a driver change on a Southampton to Garston, Merseyside, service.

On 16 June 2016, No. 66505 trundles past Whittlesea signal box with just three tanks in tow. The working is returning empty diesel fuel tanks from Ipswich loco stabling point to Lindsey, on South Humberside.

On 6 November 2014, No. 66507 heads westbound through Barnetby on one of the many coal train movements on Humberside that day.

A year earlier, on 11 June 2013, the castle is the backdrop as No. 66508 heads through Newcastle's Central station on another heavy haul coal train.

On 15 September 2020, No. 66509 is on Mendip stone duties as it approaches Fairwood Junction, near Westbury. It is working the returning empties from Avonmouth to the quarry at Whatley.

A brace of Freightliner Class 66s, with No. 66510 leading No. 66537, passes through Ipswich station's centre road on an infrastructure working from Whitemoor Yard, March, Cambridgeshire, to Parkeston, Harwich.

Electro-Motive Diesel's yard at Longport, Stoke-on-Trent, is home to several Class 66 locos on 24 March 2017. Freightliner's No. 66516 is flanked by GBRf's No. 66745 on the left and ECR's No. 66241 on the right.

Freightliner were awarded the contract to move newly built Class 387 Gatwick Express units between Derby and Bletchley depot. On 14 April 2016, No. 66518 propels unit No. 387221 round the curve towards the depot complex.

On 30 June 2021, No. 66520 passes through Severn Tunnel Junction station. The working is the return from Cardiff Wentloog FLT to Freightliner's base at Southampton.

The overbridge at the centre of Ipswich station offers this view of No. 66524 through the centre road on 2 May 2019. The liner from Felixstowe will take the Great Eastern Main Line to the London suburbs before joining the West Coast Main Line as it heads north.

On 6 September 2019, No. 66531 has taken the cross-country route across East Anglia since leaving Felixstowe on this 'liner to Crewe. It is seen passing through Melton Mowbray on 6 September 2019. The train will reach the West Coast Main Line at Nuneaton before heading north.

It's another trip down memory lane in this scene from Barnetby on 7 January 2015. Freightliner's No. 66536 is returning a rake of empty coal hoppers from Eggborough Power Station to Immingham. The traffic has ceased, the power station has closed and the much-loved semaphore signals have been replaced.

On 18 April 2018, No. 66537 approaches Grangetown Junction, having just left Wilton on the 'liner to Felixstowe. The train is seen with the industrial Teesside skyline in the background.

In addition to the facilities at Ipswich and Southampton, Freightliner's main loco servicing is carried out at their depots at Midland Road, Leeds, and Basford Hall, Crewe. On 5 July 2018, No. 66538, with No. 66614 behind, are amongst the locos stabled on the Leeds depot.

On 10 March 2020, a pair of Class 66s lead a southbound 'liner through Lincoln station. No. 66539 leads No. 66572 on this Leeds to Felixstowe service. Most container workings between Doncaster and Peterborough are routed via Lincoln, leaving express passenger services a clear run on the East Coast Main Line.

Another pair of Class 66s, this time working in top'n'tail mode, are seen at Manea, Cambridgeshire, on 26 May 2020. With No. 66413 on the rear, No. 66542 leads an infrastructure working on the short journey from Whitemoor Yard to Ely. The train will return a little later, with No. 66413 leading, having turned on the triangle of lines at Ely.

In 2018, Freightliner held the contract to operate services on behalf of Cleveland Potash from its Boulby Mine, Teesside. Most of this traffic is exported overseas through Teesport, via a dedicated terminal for potash and rock salt. On 16 April that year, No. 66551 passes through South Bank station on a loaded working from Boulby.

This photo at Derby was taken on 28 January 2016, before the station remodelling had seen the removing of these sidings. Class 66 No. 66553 has just arrived from Crewe with a short rake of former heavy haul coal hoppers now seeing use in connection with the Class 387 unit moves. The train will shortly enter Litchurch Lane Works to collect the latest unit for release.

On 29 April 2014, a single crippled coal wagon is the consist for No. 66554. The loco is on an unidentified working through the station platforms at Walsall, West Midlands.

The iconic Forth Rail Bridge sees literally hundreds of movements each day. Class 66 locomotive moves are, however, few and far between. On 5 October 2017, No. 66555 is seen heading south across the bridge returning light engine from Elgin to Millerhill Yard.

Engineering work at weekends often results in noteworthy engine moves in anticipation. On Friday afternoon 16 February 2018, news reached me of an unknown number of engines in a light loco move from Leeds Balm Road to Whitemoor Yard in readiness for engineers' trains that weekend. A dash to Newark Northgate resulted in a six-loco convoy with No. 66559 leading Nos 66599, 66593, 66414, 66518 and 66520 passing south a few minutes after my arrival.

The West Coast Main Line sees a regular procession of 'liners each weekday, both day and night. A number of destinations are served from the three major container terminals at Felixstowe, Southampton and London Gateway. On 20 May 2021, No. 66562 passes Old Linslade, near Leighton Buzzard, with an early afternoon working from Southampton to Garston, Merseyside.

On Friday 11 September 2020, Southampton's Maritime Maintenance depot was home to No. 66563. The Freightliner facility is located between the local stations of Millbrook and Redbridge. Over the next few hours a number of diesels will arrive for stabling over the forthcoming weekend.

On 15 May 2015, No. 66566 is heading for Southampton on a working from Garston. These 'liners are routed via the Northampton loop and the train is seen passing through the station's centre road.

From February 2019, Freightliner took over the operation of Mendip stone traffic from the quarries at Whatley and Merehead. The move also saw the Class 59s move from DB Cargo to Freightliner ownership. On 28 June 2021, No. 66568 leads No. 59204 on a returning 'Jumbo' working from Acton to Merehead. The pair are seen on the Westbury station avoiding line at Fairwood Junction.

A one-off wagon move on 8 February 2019 saw No. 66572 take a rake of former coal hoppers from Southampton to the yard at Leeds Hunslet. The rake is seen as it passes Burton upon Trent station.

Freightliner chose No. 66587 to receive a unique livery recognising the company's working relationship with Ocean Network Express (ONE). It was also named *As One, We Can*. ONE's choice of magenta as its house colour was inspired by the cherry blossom trees, one of Japan's national emblems. The loco is seen approaching Water Orton on 24 August 2021, with a cement working from Earles Sidings, Hope, to Walsall.

In addition to the line of Class 70s to the left, a number of Freightliner's Class 66s are on view at their maintenance facility at Midland Road, Leeds, on 22 July 2019. To the left, No. 66589 leads a line of Class 66s, with No. 66619 in the centre and No. 66420 to the right.

This is the view passing the southern end of Freightliner's Basford Hall yard at Crewe on 9 March 2018. A pair of Class 66s are on view, with No. 66596 standing in front of No. 66545.

In 2000, a new Class 66/6 subclass was built for Freightliner. These locos were built with a lower gear ratio, enabling heavier trains to be hauled, albeit at lower speeds. On 20 March 2016, No. 66603 passes Peak Forest signal box at the start of its journey from Tunstead to Brentford.

In 2017, Freightliner were in charge of the ore workings between Immingham and Scunthorpe. On 12 July that year, No. 66605 passes the now disused east signal box as it approaches Barnetby with a loaded working bound for Scunthorpe.

On 27 January 2022, No. 66613 passes Hemel Hempstead on the West Coast Main Line with a rake of empty stone hoppers. The working from Willesden will return the wagons to the quarry at Tunstead, Derbyshire.

A rainswept day in Derbyshire's Hope Valley provides the backdrop as No. 66616 returns a rake of wagons to Tunstead following weekend stabling at Barrow Hill, near Chesterfield. On 22 October 2020, the train is seen between Hope and Edale.

The Class 66/6 subclass are also to be found on workings on the Great Western Main Line to and from the stone quarries at Whatley and Merehead. On 16 June 2021, No. 66617 heads west through Reading returning empty hoppers to the Mendips.

On 13 October 2015, No. 66620 heads north through Market Harborough with the Theale to Earles Sidings working. The Class 66 is returning the empty cement wagons from the Berkshire terminal to Hope Cement Works in Derbyshire.

Sporting its Genesee & Wyoming livery, No. 66623 sits at the buffer stops at Leeds Midland Road depot on 22 July 2019. Several examples of subclass 66/6 have also been exported to work for Freightliner in Poland.

Freightliner also operates seven lower-emission examples, numbered in a subclass 66/9, namely Nos 66951 to 66957. On 9 May 2017, No. 66951 leads No. 66502 as the pair pass through Coventry station working a Crewe to Southampton 'liner.

On 13 October 2016, No. 66953 passes Market Harborough with new C2C unit No. 387303. The unit is on delivery from Derby Litchurch Lane Works to Bletchley depot ahead of mileage accumulation runs on the West Coast Main Line.

With No. 66954 exported to Poland, six examples of the subclass remain in the UK. These include No. 66957 *Stephenson Locomotive Society* seen here on Sunday 30 April 2017. The loco spent the weekend stabled on Freightliner's Southampton Maritime depot, together with a number of examples of their Class 66 and Class 70.

GB Railfreight

The company was established in 1999, several years into the post-privatisation era, unlike its main freight competitors. In its early days, an order was placed for seven Class 66s, Nos 66701 to 66707, primarily to handle its newly acquired Railtrack contract. On 13 November 2012, No. 66703 *Doncaster PSB 1981–2002* passes Worksop station level crossing on a westbound coal working.

The company's diversification since its early beginnings is nothing short of spectacular. Over twenty years since its delivery, No. 66704 *Colchester Power Signalbox*, for example, is seen on 6 August 2021 passing through Stafford on a loaded cement train from Clitheroe, Lancashire, to Avonmouth, near Bristol.

Along with other freight operators, GBRf's Class 66 locos find work annually meeting the rail network's collective effort to combat the autumn leaf fall. On 4 October 2021, for example, No. 66706 *Nene Valley Railway* (with No. 66731 on the rear) hauls a West Coast Main Line rail head treatment train (RHTT) into Nuneaton.

A number of unique brandings and liveries have been applied across GBRf's fleet of Class 66s. One of the most poignant was the application of the Ukraine flag to the loco side of No. 66708 early in 2022. On 12 April that year it is seen passing Doncaster on a Ratcliffe Power Station to Immingham empty coal working. The nameplate *Glory to Ukraine* had also been added.

Space precludes a full list of delivery and previous ownership details, if any, as GBRf's fleet expanded to a hundred locos over the next two decades. One of the most spectacular batch deliveries, however, occurred on 15 February 2016. No. 66708, then named *Jayne*, on the far left, is seen pausing alongside Worcester's Shrub Hill station whilst delivering Nos 66777, 66776, 66779 (under wraps), 66778, 66774, 66773 and 66775 from Newport Docks. An 'unwrapped' photo of No. 66779 appears later in this publication.

On 20 June 2018, No. 66709 heads south through Doncaster on an unidentified coal working. This loco was chosen to carry a unique livery marking its early contract with Mediterranean Shipping Co. (MSC). The loco side is adorned with a complete container ship and the nameplate *Sorrento*.

Another customer corporate livery was applied to No. 66711 *Sence*, in recognition of the importance of Aggregate Industries as customers of GBRf. On 26 April 2018, the loco approaches Nuneaton heading west on a working from Wellingborough to Northampton. The loco will run round its train at Whitacre Junction, on the outskirts of Birmingham, before returning to Nuneaton to head south on the West Coast Main Line.

For a number of years, the Saturday 'sand' empties between Doncaster and Peterborough were renowned for the multiple locos found on the working. Saturday 12 July 2014 was not to disappoint. On this occasion, the train was hauled by No. 66712 *Peterborough Power Signalbox*, with Nos 66732, 66718 and 66748 in tow as they pass Marholm, north of Peterborough.

On 23 June 2021, No. 66713 *Forest City* also gives a lift, this time to Colas Rail's No. 70811. With just a couple of wagons, the pair pass through Eastleigh station on a return working from Westbury.

No. 66714 *Cromer Lifeboat* is in charge of a lengthier infrastructure working on 20 July 2019. The train is passing Floriston crossing, on the Scottish Borders, having left an engineers' possession at Holywood, near Dumfries, to return the wagons to Carlisle Kingmoor yard.

On 3 May 2017, No. 66715 *Valour* is seen on one of the few remaining coal trains still operating in the UK. With a loaded working from the docks at Immingham, the rake is heading through Barnetby whilst working to the power station at Ratcliffe on Trent. The Nottingham coal-fired plant was originally commissioned in 1968 and remains operational.

On 1 December 2017, No. 66716 *Locomotive & Carriage Institution Centenary 1911–2011* approaches Clapham Junction with a rake of empty hoppers. The train is returning from Aggregates Industries (AI) at Angerstein to the AI base at Bardon Hill, Leicestershire.

Two Class 66s were to be found alongside GBRf's small maintenance facility at Peterborough on 27 August 2019. No. 66719 *Metro-Land* and No. 66731 are stabled, viewed from passing on the East Coast Main Line.

On 25 June 2020, No. 66720 passes through Water Orton station on an infrastructure working taking empty wagons from the yard at Eastleigh to the Leicestershire quarry at Mountsorrel for loading. The loco is about to lose its distinctive 'rainbow' livery as it has entered Eastleigh for a repaint.

Another distinct livery is carried by No. 66721 *Harry Beck*. The loco's side highlights the familiar London Underground tube map, as well as promoting the London Transport Museum. The loco is seen on 23 September 2016 entering the sidings at the former South London depot at Hither Green.

Reflecting an era when GBRf was under the ownership of the First Group, No. 66724 *Drax Power Station* passes Peterborough on 1 November 2012. The loco is heading north on the gypsum service from Mountfield to West Burton Power Station.

Sister loco No. 66725 *Sunderland*, also in First Group livery, is seen on 9 September 2012. On this occasion, the working is a one-off taking a rake of new 'Greenergy' branded tanks to the oil terminal at Immingham. The train has just passed through Newark's Castle station.

The working partnership with Maritime Transport was recognised by the unveiling of No. 66727 in that company's corporate style and naming, Maritime One. It was to be Ministry of Defence action for the loco on 20 February 2020. The loco is passing King's Sutton on a short MOD rake working from Bicester to Kineton.

For many years, the infrastructure service linking Bescot and Toton yards was in the hands of DB Cargo. Now in the hands of GBRf, their No. 66729 *Derby County* passes Water Orton on 20 April 2021. The train is conveying point carriers to Toton.

One of GBRf's most popular decisions in recent years was to outshop No. 66731 with Thank you NHS vinyls on the loco's bodyside. The company also seized the opportunity to name the loco *Capt. Tom Moore – A True British Inspiration*. It is seen at Nuneaton on 13 June 2022, taking a single crippled box wagon from Knighton Junction, near Leicester, to Bescot Yard.

The diversity of duties of GBRf's Class 66s is further evidenced in this photo of No. 66733 *Cambridge PSB* on 16 October 2021. The loco, together with No. 73969, has just arrived at Fort William at the head of that town's portion of the Highland Sleeper from London Euston. The Class 66 has worked the train from Edinburgh.

On 1 December 2011, No. 66734 is seen heading south through Nuneaton on a working from Carlisle to Whitemoor. This loco is the only casualty of the company's Class 66 fleet, having been written off a year later following a derailment between Corrour and Tulloch on the West Highland Line in June 2012. Unusually, the TOPS number, 66734, has since been reused.

GBRf's fleet expansion continued on 21 December 2012. Loco No. 66735 had been despatched to Dollands Moor to collect Nos 66747, 66748 and 66749, which had arrived in the UK through the Channel Tunnel the night before. The four-loco convoy then headed north to the Midland Railway Centre, Butterley. The convoy is captured pausing at Kettering station.

Another Class 66 is found on sleeper duties on 13 October 2021. This time, loco No. 66737 *Lesia* is in charge of the Inverness portion of the Highland Sleeper. It has just terminated at Inverness, paired with No. 73967.

The only regular freight turn on Scotland's West Highland Line is also in the hands of GBRf Class 66s. It operates loaded alumina trains from North Blyth, Northumberland, to the Lochaber aluminium smelter just outside Fort William. On 18 October 2021, the southbound working is seen passing through Upper Tyndrum behind No. 66741 *Swanage Railway*.

Two of GBRf's fleet, Nos 66743 and 66746, were given Belmond Royal Scotsman branding specifically for working these prestigious passenger services, chiefly in Scotland. On 12 October 2021, No. 66743 is seen crossing Tomatin viaduct, on the Highland Main Line, as it heads south to Aviemore.

Another GBRf-operated RHTT is seen heading south through Nuneaton on 30 September 2019. With No. 66745 *Modern Railways – The First 50 Years* and No. 66733 operating in top'n'tail formation, the water jets can clearly be seen.

GBRf's second Belmond-liveried loco, No. 66746, is a long way from Scottish passenger duties on 19 May 2020. It is crossing the Fens approaching Manea at the head of a Felixstowe to Birch Coppice intermodal working.

Another GBRf partnership was recognised by No. 66747's emergence in the colours of Newell & Wright. The Sheffield-based container handlers have a rail-accessed base at Masborough, Rotherham. The loco also sports the name *Made in Sheffield* as it passes Water Orton on 7 July 2020, working a rake of empty hoppers from Tinsley, Sheffield, to Coton Hill, near Shrewsbury.

On 5 June 2021, weekend engineering work on the southern end of the ECML saw No. 66748 *West Burton 50* diverted via the Midland Main Line heading north from London. With two Southeastern units in tow, the move from Slade Green depot to Doncaster has just emerged from Ampthill tunnels, Bedfordshire.

Recent UK arrival No. 66749 is still in its original grey colour scheme on 22 June 2013. The loco is seen passing Hillam Gates crossing on a light engine move from Derby to Tyne Dock.

Similarly, No. 66751 retains a plain blue livery when seen at Derby on 8 November 2013. The loco has a Beacon Rail logo under the cab window, and carries the words 'Maintained by ElectroMotive' on its bodyside.

GBRf's UK expansion now extends to South Wales. On 30 June 2021, No. 66752 *The Hoosier State* heads west through Cardiff Central returning a rake of vans from Tilbury Docks to Trostre, near Llanelli.

On 30 March 2017, No. 66753 *EMD Roberts Road* passes through Bescot Stadium station platforms as it heads towards the West Midlands yard. The loco is returning with a rake of side tipper wagons following loading at Stud Farm quarry, in Leicestershire.

On 30 August 2017, No. 66756 *Royal Corps of Signals* has just passed the signal box and station level crossing at March, Cambridgeshire. The train now heads westbound through the station on an intermodal working from Felixstowe to Doncaster.

Since 2017, GBRf has handled the china clay flow from Antwerp Docks to the Caledonian paper mill at Irvine, Ayrshire. The wagons arrive in the UK via the Channel Tunnel before commencing their journey to Scotland via the WCML. On 18 October 2017, No. 66758 *The Pavior* has just left Carlisle Yard with the northbound loaded working and is approaching the Scottish border at Floriston.

On 21 May 2015, No. 66760 *David Gordon Harris* approaches Clapham Junction station with a rake of empty box wagons, being returned to Bardon Hill, Leicestershire.

Another 2015 GBRf working sees No. 66761 head north through Stafford on 27 March that year. The Class 66 is returning a rake of hoppers from the now closed Ironbridge Power Station to the sidings at Tuebrook, Merseyside.

On 12 October 2016, No. 66762 disturbs the tranquility of the East Anglian countryside. The loco is seen coming off the now truncated line that once led to Dereham and Swaffham. The train is carrying loaded silica sand from nearby Middleton Towers, today's terminus on this branch, to Goole Glass Works, North Humberside.

It was another wet afternoon in Leeds on 9 March 2016. No. 66764 heads westbound through the station platform with a return aggregates working from Dairycoates, Hull. The empty wagons are returning to the Tilcon site at Rylstone, on the former branch line from Skipton to Grassington, in North Yorkshire.

On 22 July 2021, No. 66765 is seen passing through Coalville on a rake of empty wagons returning from unloading at Tinsley, Sheffield. They are close to journey's end at the nearby quarry sidings at Bardon Hill.

On 13 August 2021, No. 66767 *King's Cross PSB 1971–2021* was a late loco substitution for a very different working. As a late replacement for a more compatible partner for No. 56081, the Class 66 brings up the rear of this new unit delivery move from Derby to Wembley. The train is seen on the Marston Vale branch as it passes Lidlington, Bedfordshire.

Shortly after commencing its journey from Felixstowe, No. 66770 is seen at Westerfield, on the branch between the docks and Ipswich. The train is destined for Hams Hall terminal in the West Midlands. In 2022, GBRf celebrated twenty years of operations out of the Port of Felixstowe.

A much more recent GBRf contract sees the movement of spoil in connection with the HS2 construction work. These trains are operated between Willesden and Barrington, Cambridgeshire. On 19 August 2021, No. 66771 *Amanda* comes off the short Barrington branch spur and heads through Foxton station returning the empty wagons to London.

A bond of a different kind led to GBRf's naming and branding of No. 66775, following several GBRf employees experiencing four days aboard Type 23 Frigate, F231 HMS *Argyll*. No. 66775 became F231, and now carries the Frigate's name, *HMS Argyll*. On 26 April 2022, the loco passes Water Orton on a light engine move to Eastleigh Yard.

On 19 July 2021, No. 66778 *Cambois Depot 25 Years* heads north through Leicester station taking a rake of empty box wagons from Wellingborough Yard for loading at the Derbyshire quarry at Tunstead.

Earlier in this publication newly built No. 66779, literally under wraps, was seen passing through Worcester on delivery in February 2016. Three months later, the loco, still in pristine condition, is on static display inside the National Railway Museum, York. The loco's *Evening Star* nameplate had also been unveiled.

On 9 June 2021, No. 66780 is seen approaching Peterborough on a Yorkshire-bound intermodal working from Felixstowe. The loco now carries a unique livery, reflecting the importance to GBRf of their customer, Cemex. The loco was also named *The Cemex Express*.

In 2018, No. 66783 was chosen by GBRf to carry a special livery marking the working partnership with the waste management company Biffa. I believe the loco's name, *The Flying Dustman*, was the winning entry of a naming competition amongst the Biffa workforce. On 19 November 2021, No. 66783 is seen at Harrowden Junction, north of Wellingborough, working the empty cement wagons back to Ketton from St Pancras Churchyard Sidings.

GBRf operate aggregate traffic between Tunstead Quarry and Small Heath, West Midlands. On 21 October 2020, No. 66785 is approaching Great Rocks Junction and about to enter the Tunstead complex with the returning empties from Small Heath.

The GBRf fleet expansion had continued with ten, randomly numbered, acquisitions from DB Cargo, which were renumbered 66780 to 66789. On 2 August 2019, No. 66787, which was previously DB Cargo's No. 66184, passes under the station footbridge at Mexborough, South Yorkshire. The loco is working a Toton to Doncaster infrastructure service.

The last of this batch of ten, No. 66789, was chosen to receive a modern equivalent of the BR Blue Large Logo livery, together with the nameplate *British Rail 1948–1997*. The loco passes Leicester station on 28 March 2018 on Theale to Immingham with a rake of empty fuel tanks.

Dating back to 2002, No. 66790 was an import from Sweden in June 2019 for conversion to UK standards by Electro-Motive Diesel. On 16 March 2022, the loco was seen passing through Gainsborough Lea Road on a Whitemoor to Doncaster engineers' train.

The second acquisition from Sweden, to become GBRf's No. 66791, was also converted by Electro-Motive at Longport, Stoke-on-Trent. On 13 March 2020, it was being shunted around the Longport complex by resident shunter No. 08784.

GBRf's No. 66794 was chosen to receive BR Railfreight triple grey livery, along with Petroleum sector decals. Additionally, the loco was named *Steve Hannam*, to mark the retiring railwayman's contribution to both Class 59 and 66 locos for almost half a century. The loco passes Bedford while returning an empty rake from Acton Yard to Bardon Hill Quarry.

On 27 July 2021, No. 66795 heads south on the Midland Main Line on the loaded aggregates working from Peak Forest to Limbury Road, Luton. The train is passing Oakley, just north of Bedford.

GBRf's No. 66799 was put into UK use at the beginning of 2022. On 25 April it hitched a lift on the front of a Felixstowe-bound intermodal from Birch Coppice, in the West Midlands. The loco is seen together with No. 66742 as it passes Leicester. I believe No. 66799 was removed from this working at Peterborough.

Colas Rail

The final operator of Class 66 locos is Colas Rail. They currently have a small fleet of five locos, numbered 66846 to 66850. These were built in 2004 and formerly owned by Freightliner, numbered 66573 to 66577. They are regularly found at the southern bases of Eastleigh and Hoo Junction. On 11 September 2020, Nos 66846, 66850 and 66849 are waiting to leave Eastleigh on a light engine move to Hoo. Most of the company's work in the area is on infrastructure workings on behalf of Network Rail.

One of the few regular freight trains operated by Colas is on behalf of Kronospan. Colas Rail handles timber traffic to and from their plant at Chirk, near Wrexham, with the majority of workings between there and Carlisle. On 7 July 2012, No. 66848 is seen about to run round the returning empties at Shrewsbury station.

The Colas Class 66s make occasional visits to the company's Rugby depot. On 15 February 2013, for example, No. 66849 *Wylam Dilly* can be seen in the yard surrounded by a variety of Colas Rail's On-track plant.

For several years, the company provided the diesel power for the engineers' working between the yards at Hoo Junction and Whitemoor. On 30 August 2017, the train, hauled by No. 66850 *David Maidment OBE*, has reached March, Cambridgeshire, and is crossing over to gain access to the short spur into Whitemoor Yard.